PIRATES

PIRATES

ODYSSEYS

RACHAEL HANEL

CREATIVE EDUCATION · CREATIVE PAPERBACKS

Published by Creative Education and Creative Paperbacks
P.O. Box 227, Mankato, Minnesota 56002
Creative Education and Creative Paperbacks
are imprints of The Creative Company
www.thecreativecompany.us

Design by Graham Morgan
Art direction by Tom Morgan

Images by Getty Images/Bristol Archives, 32–33, Heritage Images, 2, 26; Unsplash/Alonso Reyes, 62, DaYsO, 6, iAko Randrianarivelo, 70–71, Tanner Mardis, 22; Wikimedia Commons/Benjamin Cole[2], 58, Chatterton, E. Keble (Edward Keble), 18, Funk Dooby, 24, Geo. S. Harris and Sons, 50, Howard Pyle, cover, 44, 53, Jean Leon Gerome Ferris, 11, 12, Laureys a Castro, 64, Lenish Namath, 48, Lieve Verschuier, 4–5, Maurits van Nassau (Maurice of Orange, 1567–1625) / Melander, secretaris, 17, Orem, 54, public domain, 37, The Smithsonian Institution, 29, Theodore Scott, 75, Thomas Luny, 66, U.S. Navy, 69, WarX, edited by Manuel Strehl, 8

Library of Congress Cataloging-in-Publication Data
Names: Hanel, Rachael author
Title: Pirates / Rachael Hanel.
Other titles: Pirates (Odysseys in combat)
Description: Mankato, Minnesota : Creative Education and Creative Paperbacks, [2026] | Series: Odysseys in combat | Includes bibliographical references and index. | Audience term: juvenile | Audience: Ages 12-15 Creative Education and Creative Paperbacks | Audience: Grades 7–9 Creative Education and Creative Paperbacks | Summary: Prepare for battle alongside pirates sailing the lawless high seas by learning about their weapons, combat skills, and reasons for fighting. Includes a glossary, sidebars, index, and further resources— Provided by publisher.
Identifiers: LCCN 2024049675 (print) | LCCN 2024049676 (ebook) | ISBN 9798889896494 library binding | ISBN 9781682778159 paperback | ISBN 9798889897293 ebook
Classification: LCC G535 .H263 2026 (print) | LCC G535 (ebook) | DDC 910.4/5—dc23/eng/20241114
LC record available at https://lccn.loc.gov/2024049675
LC ebook record available at https://lccn.loc.gov/2024049676

Printed in India

A treasure chest is an enduring pirate symbol.

CONTENTS

Introduction

From the beginning of time, wherever groups of people have lived together, they have also fought among themselves. Some have fought for control of basic necessities—food, water, and shelter—or territory. Others have been spurred to fight by religious differences. Still others have fought solely for sport. Throughout the ages, some fighters have taken up arms willingly. Others have been forced into battle. For all, however, the ultimate goal has always been victory.

OPPOSITE: Pirates often used the skull and crossbones symbol to instill fear in sailing crews they met on the high seas.

OPPOSITE Since the 1700s, storytellers have glamorized the lives of pirates such as the colorful Captain Kidd.

Pirates engaged in battle for one primary purpose: to make themselves rich through stolen treasures. For centuries, pirates terrorized high-seas captains and crews the world over. They slashed and fought their way onto ships that carried gold, silver, and other valuables, killing or injuring crews that did not hand over treasures willingly. With their newfound riches, pirates docked at various ports around the globe, relaxing and spending their money on alcohol, gambling, and women. Their wild and rough behavior earned them notorious reputations. Although faster ships and improved communication in the 19th century finally lessened the dangers at sea, pirates still exist today. And pirate stories remain popular, captivating people's imaginations and appealing to their curiosities.

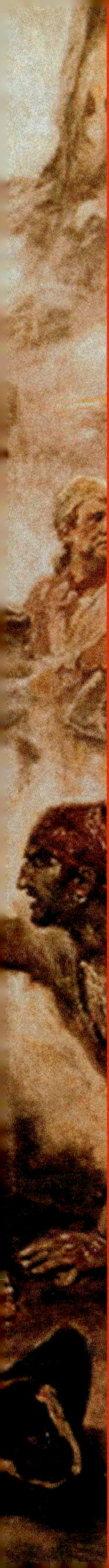

An Age of Treason

Pirates dominated the seas in the 16th, 17th, and 18th centuries, a time when sea travel spanned the globe. For centuries before the rise of pirates, the wide expanses of the Atlantic and Pacific oceans remained uncharted, as the technology to navigate waters and set sail for months at a time had not yet emerged. Sea travel during that time took place only in smaller areas such as the North Sea, the Mediterranean Sea, the seas off the coast of Asia, and waters near Africa.

OPPOSITE: Blackbeard and other notorious pirates never gave up without a fight.

Through the beginning of the 16th century, trade was limited to Europe, Asia, and Africa. Europeans and Asians exchanged goods and ideas using various land trade routes. Europeans and Africans ventured across the relatively small Mediterranean Sea to trade with each other. They spun the beginnings of a web that soon connected most of the Eastern Hemisphere. This trade brought riches to powerful merchants and the kings and queens who ruled over them. With this wealth came the desire for more wealth. In the 1400s, the technology emerged that allowed for wider sea travel. Scientists discovered that the sun and stars could serve as guides across the oceans. Tools such as the compass and **astrolabe** aided navigation, and ships became larger, faster, and sturdier. With that, the **Age of Discovery** began.

Europeans soon crossed the Atlantic Ocean and discovered what would become known as America.

They christened the territory the "New World." Europeans saw in the vast wilderness of the New World the potential for domestication and crop development. They also discovered precious metals such as gold and silver in these lands. Suddenly, they had before them new ways to add to their riches. It didn't take long before wealth and treasures flowed back and forth between the Eastern and Western hemispheres aboard large ships. It also didn't take long before criminally minded men identified opportunities to become wealthy themselves through robbery and theft.

Pirates during the Age of Discovery emerged from the fringes of society. They were outcasts, men who didn't fit into the rigid class structures of the day. Some were lifelong criminals. Others had served at sea in a military capacity but disliked the strict structure and

OPPOSITE Privateers carried government documents to prove they were operating under the law.

formal chain of command. Some simply embraced a sense of adventure. These misfits found strength in meeting others like themselves and bonded together, forming pirate crews and **fleets**.

At times, governments officially sanctioned piracy. During the Age of Discovery, European nations routinely fought against each other. Governments eager to attack the enemy employed pirates, in this case called privateers, to seize enemy ships and rob them of their treasures. The privateer shared some of the

n de moor ende
compaignie,

Maurits bijder gratie Godts etc. Saluyt, Alsoo ons Joha
Moor, woonachtich binnen Vlissingen ende Compaignie te kennen
gegeven hebben, datse nu eenen langhen tijt van jaren, hebben ge
negotieert opde Westindien, ende voornemelijcken op ende inde ri
van Amazones. Wajapota. ende andere revieren ende quartieren
ontrent gelegen, Ende aldaer gedurich eenighe schepen ende ja
als mede eenich volck, aldaer aen lant gehouden hebben, ende b
noch houdende sijn, tot gerief vande huysen aldaer getimmert,
bij haer geemployeert worden, ten deele omme de selve quartier
te landewaerts breeder te ondersoecken ende tontdecken, als mede,
omme de lantbouwerye (bij haer inde voorss quartieren aengevan
waer te nemen ende tegens d'invasie van quaetwillighe te beschu
Ende dat sij bevreesen teenigher tijt vande Spaignaerden overval
ende in haere handelinghe verhindert te sullen worden, ende ons ove
sulcx versocht wij wilden haer consenteren, dat sij (in dien gevall
hun souden moghen ter weere stellen, tegens den Spaignaerden
alsulcke andere, die hun in desen souden willen beschadighen oft
verhinderen, Soo ist, dat wij aenmerckende, den intere
ende grooten ondienst, die dese landen daerbij souden hebben te l
den voorss De moor ende Compaignie volcomelijck geauthoriseert h
door haere bevelhebberen so te water als te lande, met de
wapenen te moghen wederstaen, tot haere defentie, allen de
gheenen die hun inde voorss quartieren, hoo in haere handelinghe
eenighe schade ofte belet soude willen aendoen, als omme t
verhinderen, dat haer diergelijcke overlast niet meer en gesch
Consenteren hun (in dien gevalle) datse den Spaignaerden
sijne ondersaten, oft anderen haeren vianden, sullen moghen s
achterhalen ende aentasten opde custen van Brasilien ende Westjn
daer sij de selve sullen comen vinden, omme te verhalen haer
schaden die sij alreede van hun souden moghen geleden hebben,
Behoudelick nochtans, dat so wanneer sij eenighe goederen inde
nemeren ende uijt oorsaecke voorss, souden comen te verover
de selve gehouden sullen wesen te brenghen in dese Landen, omm
als goeden buyt ende prinse bijde Raden vanden Admiraliteyt, da
die hunne gebrocht sijn ofte daerinne anders gedisponeert te wer
naer behooren, Ende hun voorts te reguleren, achtervolgende
den Placcaten van Ho: Mo: heeren Staten generael deser

A Long History

Although piracy reached its height during the Age of Discovery, pirates existed and terrorized the seas long before this. Wherever sea exploration took place, pirates were sure to be part of that culture. Ancient peoples such as the Greeks, Phoenicians, and Romans were known as expert seafarers, exploring the far reaches of the Mediterranean Sea, and had to deal with pirate attacks. The Vikings of the 9th, 10th, and 11th centuries were also known for their excellent navigational skills. The Vikings often participated in piracy, sailing from Scandinavia down the western coast of Europe and attacking ships and towns along the shores.

treasure with the government officials who had hired him. Privateers possessed letters of marque written by government officials. These letters cleared the privateer of any wrongdoing in case he was captured. Although officially sanctioned, privateers often acted in brutal ways, and there was little difference between them and ordinary pirates.

Pirates and privateers the world over used violence to get their way. They boarded ships and threatened crew members until they turned over their treasures. Only the bravest and most well-equipped sailors dared to stop them, so pirates were able to **plunder** and steal with little consequence.

Despite occasional riches, however, the life of a pirate was not easy. Pirates stayed at sea for weeks, months, and even years at a time, living with anywhere from a few to a dozen to hundreds of other pirates. During the rare times when they weren't sailing, they might return home or else make a new home at a port. Aboard ship, pirates slept wherever they could find space. They might use a sack of wheat as a makeshift bed. Usually, only the captain had his own private quarters.

Sailing was hard work, and each pirate had a different duty. The captain led the crew. The quartermaster, who was second in command behind the captain, often mediated fights among pirates. He was also supposed to be the first to board a target ship. First and second mates assisted the captain and quartermaster. The sailing master kept charge of the navigation and sails, while the boatswain supervised the maintenance and day-to-day operations of the ship. The gunner knew how to use weapons to their best efficiency during an attack. Carpenters helped repair ships. Pirates who were injured and unable to perform other duties might serve as cooks. Because of the demanding tasks they had to perform, pirates tended to be young. Older men usually

Hoisting heavy cloth sails required brute strength.

did not have the physical strength and stamina required to carry out such jobs.

The seas often treated pirates roughly. Strong storms rocked ships and sometimes even sank them. On deck, pirates breathed in fresh air but were also exposed to the elements. Whipping winds pelted crews with rain and salty ocean spray, making them wet, cold, and uncomfortable. If they went below deck, the space was dark, cramped, and damp, with dirty water sloshing back and forth and rats scurrying about.

Pirates faced a host of illnesses and injuries aboard a ship. They frequently lost limbs and eyes in battles or in onboard accidents with sails and rigs. Fresh food was nonexistent. Instead, pirates lived off biscuits or heavily salted meat. They enjoyed fresh meat only if they had just left port. Sometimes they caught sea turtles and cooked

Talk Like a Pirate Day

September 19 marks International Talk Like a Pirate Day. Pirates had a language all their own, with exclamations such as "Aaaargh!" and phrases such as "Avast, ye mateys!" (which means "Stop!" or "Hey!"). Two American men—John Baur and Mark Summers—injected this type of talk into their everyday conversations just for fun. In 1995, they established Talk Like a Pirate Day. The two marked the day quietly themselves for several years, but in 2002, the holiday received international media attention. The day now has its own website and many devotees.

them for food. Because fresh fruits and vegetables were not available, pirates often suffered from **scurvy**.

Pirates escaped their harsh life once they docked on land. Here they lived for the moment and enjoyed carefree times. Contrary to popular belief, pirates did not bury their treasure. Instead, they spent it freely on alcohol, gambling, and women at the nearest port. Some pirates married many times over at different ports. While on land, pirates also had to attend to business, stocking up on food and medicine and making needed repairs to their ships.

With their reputation for violence and coarseness, pirates terrified most people. Even government officials who sanctioned privateers viewed them as nuisances because of their coarse manners.

Fighting, Pirate Style

A pirate's most important sidekick was his weapon. Whether he wielded a sword or a gun, a pirate relied heavily upon his weapon, for without it, he was unlikely to sail away with treasure. A glinting blade or the threat of a bullet made a pirate's opponent much more apt to yield to his wishes.

OPPOSITE: Pirates often swarmed ships that didn't immediately surrender, and fierce hand-to-hand combat ensued.

Pirates traversing the seas in the early years of the **Age of Sail** depended heavily upon a short, thick sword known as a cutlass. The cutlass featured a slightly curved blade, with a comfortable handle at the end, usually wrapped in leather. It not only inflicted bodily harm but also was used for routine onboard jobs, such as cutting through thick rope or canvas. Its compact size allowed for use in cramped quarters below deck and also made it easy to use with little training. Pirates also used a dagger, which had a considerably shorter blade than a cutlass. Pirates used their daggers as secondary weapons for close hand-to-hand battle. Like the cutlass, the dagger also was used for onboard jobs, such as cutting food.

When guns first emerged around the 16th century, pirates quickly seized upon these weapons. A gun allowed a pirate to fire from a longer range, keeping him safer than in hand-to-hand combat. Pirates used a musket, which is a longer gun, and often kept a pistol, a shorter gun, at their sides. Guns were less reliable than swords because they were cumbersome. A pirate had to reload the gun after each shot in a time-consuming process. Aiming

Flintlock mechanism on a musket

correctly was also difficult as ships bobbed up and down in the water. Because of these difficulties, many pirates felt more comfortable battling with their cutlasses and daggers, even though they had to get closer to their opponents in order to use these weapons.

Advanced pirate weaponry included cannons and makeshift bombs. Pirates used cannons to fire warning shots when approaching a target ship in hopes that the ship would surrender without a fight. Cannons were difficult to use, though. They were heavy, and the strength of several men was required to hoist the massive cannonball into the weapon and fire it. Rather than using a cannon, pirates sometimes hollowed out a much-smaller cannonball, filled it with gunpowder, lit a fuse, and threw it onto an enemy ship. However, making such a bomb required precise skill, and sometimes the venture backfired, killing or injuring the pirate instead of the intended victim.

Sometimes pirates put a new spin on old weapons. Ancient pirates sometimes mixed oil and tar, set it on fire, and threw it at enemy ships. This was known as Greek fire. During the Age of Sail, pirates used a variation of Greek fire. They filled clay pots with stinky mixtures of sulfur, rotten food, and dead animals, then lit these on fire and hurled them onto ships so that the choking smoke would disable their victims.

The Wild, Wild Port

Port Royal, Jamaica, was a favorite hangout for pirates in the 17th century. The British first built a fort there in 1655, and within a few years, hundreds of houses and businesses surrounded the fort. Since the British lacked troops for the fort's defense, they instead employed pirates to guard the city. Port Royal quickly became known as the wildest, roughest city in the world. Drinking and gambling were popular there—there was a tavern for every 10 people. An earthquake devastated the city in 1692, but it was rebuilt over the following decades and is now a small fishing village that attracts tourists year-round.

A (   )

he Dockyard –
Port Royal. Jamaica

Pirates also used a primitive form of a **hand grenade**. They filled a glass bottle with gunpowder, lit it, and threw it aboard an enemy ship. Sometimes, if pirates ran out of ammunition, they blasted gold coins out of cannons at their victims. The force behind the propulsion lodged the coins in the victims, killing or injuring them. Once aboard the ship, pirates used scalpels to retrieve the pieces.

Pirates acquired their weapons while on land, either at home before they sailed for sea or while docked at various ports. Expert swordsmen and blacksmiths created

PIRATES USED FLAGS WITH FRIGHTENING PICTURES, SUCH AS THE SKULL AND CROSSBONES . . . IN HOPES OF SCARING A SHIP'S CREW INTO SURRENDER.

the weapons. Sometimes, too, pirates gathered weapons as part of the **booty** when they captured a ship. A good weapon was nearly as valuable as money.

Of course, pirates would have no chance of success without a good ship to move them swiftly across the seas. Pirates used faster, sleeker, and smaller ships than those they were trying to capture. **Merchant ships** carrying goods and treasures were large and bulky in order to accommodate cargo, and they moved slowly. In contrast, pirates used shallow ships—that is, little of the ship stayed below the waterline. This helped captains quickly navigate waters. Asian pirates used a junk—a flat-bottomed boat with bamboo rods holding up the sails. At times, pirates used captured ships and refit them to become part of their own fleet.

Like other ships on the high seas, pirate ships used sails to catch the wind and move. The heavy, tough canvas sails were designed to withstand strong storms and pelting rain. In times of little wind, or when pirates hoped to quickly gain on an enemy, crews used oars to sluice through the water.

Flags served an important purpose on pirate ships. All ships—whether pirate or regular trade ships—used flags, which designated the country they represented.

Seize Caesar!

The noted Roman emperor Julius Caesar (100–44 BCE) was once captured by pirates. Before he became ruler of the **Roman Empire**, Caesar served in the Roman navy and worked to fight pirates in Cilicia (present-day Turkey). During a battle in 78 BCE, a group of especially cruel pirates captured Caesar and held him hostage. They forced the Romans to pay a large ransom to reclaim him. The pirates continued to dominate the waters until 67 BCE, when the Romans finally defeated them, killing 10,000 and capturing twice as many.

Sometimes pirate ships flew the flag of a different country, which tricked passing ships into thinking the pirate ship was friendly. Other times, pirates used flags with frightening pictures, such as the skull and crossbones—known as the "Jolly Roger"—in hopes of scaring a ship's crew into surrender.

Pirates acted flamboyantly and wildly, and their manner of dress reflected these characteristics. North American and Mediterranean pirates often wore lavish, long coats over leggings or stockings and some type of boot or buckled shoe. In warmer waters, such as the south Asian seas, pirates wore less clothing, often sleeveless tunics and short skirts. Always, though, pirates' clothes were colorful.

Pirates the world over wore something on their heads, either a hat or bandana, to protect themselves from the weather. Those who lost an eye to injury wore an eye patch, while those who had lost a limb used a wooden leg or hook arm. Many pirates also wore gold hoop earrings. Some historians believe one explanation for this is that if a pirate died in battle, the gold would provide enough money for a funeral. Another explanation is that pressure on the earlobe prevented a pirate from becoming seasick. Pirates also wore the jewelry they found in looted treasures as a way of flaunting their successes.

Clashing on the High Seas

Each day aboard a pirate ship brought the chance of spotting a merchant ship far off on the horizon. Like a mirage, the ship would at first shimmer in the distance. Taking turns looking through **spyglasses**, the pirates would verify that, yes, they had a target before them. It was time to get to work.

Pirates rushed about the ship. They gathered their weapons and secured them at their sides. The pirates who were responsible for the sails and rigging hoisted them to catch even the slightest breeze and gain speed on the target. They raised a flag of the same color as the target ship; that way, they could get close to the ship by appearing to be an ally from the same country. Sometimes pirates even wore disguises, dressing as women or naval captains to appear as innocent passengers or officials.

As the pirates closed in on their target, they raised a white flag to signal to the other ship's crew that they should surrender peacefully. If the crew refused to do so, the pirates hoisted a red flag, which announced one thing: bloodshed and violence awaited. If the opposing ship still did not surrender, the pirate ship might use cannons or bombs to disable the ship or injure its crew.

Then, coming up directly behind the target ship, the pirates rammed it and tried to disable its rudder so it couldn't be steered. At the same time, the pirates had to be very careful to keep the vessel intact—if it sank, its treasures would go down with it. With the captured ship still in one piece, the pirates could also eventually make it part of their fleet.

If it became clear that the crew of the merchant ship was not going to surrender, pirates maneuvered their ship alongside the merchant ship and cast long ropes with hooks on the ends onto the ship's deck. Using the ropes, they pulled the ship closer. They might also use grenades to set the merchant ship's sails on fire or create billowing smoke to confuse the crew. Then, pirates swarmed aboard the ship. Within moments, pirates and crew members were engaged in brutal hand-to-hand combat. Crews

Somali Pirates

In 2005, Somali pirates attacked a United Nations ship carrying food for victims of the tsunami that struck Southeast Asia and the eastern coast of Africa in December 2004. The MV *Semlow* was carrying 850 tons (771 metric tons) of rice, but the pirates were not interested in its cargo. Instead, they held the 10 crew members—8 Kenyans, a Tanzanian, and the Sri Lankan captain—hostage, demanding $500,000 in ransom from international authorities. Details of the negotiations remain unclear, but the crew was released after nearly 100 days, and the ship's cargo was left intact.

that surrendered right away and allowed the pirates to take their cargo might escape unharmed, although they might be forced to serve as pirates or perform other duties, such as carpentry, aboard the pirate ship.

Braver crews, though rare, might try to thwart the would-be thieves. At times, they greased the deck so pirates would slip and fall. Some scattered peas or other food to make walking difficult. If the crew members were successful, they would stave off the attack and kill or injure the pirates. Trying to outwit or outfight pirates was a risky business, though. If the crew failed and the

Walk the Plank

In countless novels and movies, a favorite punishment doled out by pirates for any major breach of the pirate code is "walking the plank." Pirates force their offending comrades to walk out onto a plank over the sea, usually blindfolded. Their final step is down into the ocean, where they drown or are eaten by sharks. Although there is little documented evidence of real pirates doing this, the practice is thought to have originated with Greek pirates, who forced their Roman captives to suffer the punishment.

pirates captured the ship anyway, a violent and torturous death awaited those who had opposed the pirates.

Not all pirate attacks involved boarding ships. Occasionally, pirates disembarked and raided villages and towns along coasts. Some larger pirate boats, especially Asian junks, held a number of smaller rowboats that pirates used to get to land. On land, pirates went from building to building looking for goods and killing or injuring anyone who got in their way.

Whether by land or sea, pirate attacks were often successful, despite the fact that pirates did not undertake years of formal training. Instead, they brought a variety of life experiences with them. Some were already good fighters because they had prior military experience and knew how to handle deadly weapons. Other pirates learned their skills on the job. They simply picked up a dagger or pistol and practiced with it. More experienced pirates might help them in their training.

Surprisingly, pirates operated under a **democratic** system that stressed equality at a time when most European societies were ruled by powerful kings and queens. Many pirates had already experienced the strict, tough discipline found aboard naval ships in the military, where captains, who represented an elite class, treated their crews poorly. Pirates who had served on these crews disliked the rigid structure and class differences. In their new lives, they sought a more equitable arrangement. Rather than assigning a leader, a crew voted upon a captain. He was generally a man whom all trusted and respected.

Although the captain was the highest-ranking member aboard a ship, he did not make all the decisions. Together, all the pirates on the ship or within the fleet decided where to search for merchant ships and how the treasure should be divided. The captain received the most shares, while other important men on the ship, including the quartermaster, the master gunner, and the carpenter, received more shares than the regular crew. Crews agreed upon rules for everyone, such as what constituted intolerable behavior and how much a pirate should be compensated for an injury, and determined punishments.

Should a pirate break a rule, the crew decided by vote how to punish him. A pirate who always argued,

A deserted island

threatened violence, or tried to take all of the booty was considered dangerous. Punishment might include chaining the offender in leg irons or whipping him until the captain felt he had learned his lesson. A more serious punishment was keelhauling, in which the victim was tied to the end of a rope and dragged under the ship from one side to the other. A pirate usually died from this punishment, as either his lungs filled with water and he drowned, or he struck the bottom of the ship, which was covered in rough barnacles that cut his skin and caused him to bleed profusely.

The worst punishment—doled out for an offense such as stealing from a fellow pirate or excessive fighting—involved abandonment on a deserted island. A crew might leave the pirate just a weapon and some drinking water. The water supply didn't last long, and the marooned pirate usually died a slow death from starvation and dehydration.

The Terrifying, Unique, and Successful

The flamboyant and violent ways in which pirates acted earned them superior status among their peers. During their lifetimes, certain pirates became well known because of their successes and battles. Stories of their exploits were passed down from generation to generation until the stories took on lives of their own.

OPPOSITE: Henry Morgan, a privateer who operated largely in the Caribbean, is today known as the face of a brand of spiced rum.

One of the most famous pirates ever to terrorize Atlantic waters was Blackbeard (c. 1680–1718). Blackbeard dominated the waters off the North American coast and the Caribbean Sea from 1716 to 1718. Born Edward Teach (or Thatch) in Bristol, England, Blackbeard adopted his nickname to reflect his intimidating look—jet black hair and a long, black beard, which he separated into braids. He placed pieces of burning rope under his hat so that smoke billowed out and added to his menacing appearance. Blackbeard was said to have occasionally killed members of his own crew just to prove his dominance.

Blackbeard ambushed ships in the Caribbean and the Gulf of Mexico and took his looted treasures back to his favorite hideout on Ocracoke Island off the coast of North Carolina. He then sold his goods, usually at lower prices than otherwise offered. The American colonists

Blackbeard

Black Bart's Flag

Captain Bartholomew "Black Bart" Roberts (1682–1722), an 18th-century pirate from Wales, designed a special flag to send a sinister message to his enemies, the governors of Barbados and Martinique, who were constantly trying to capture him. The flag featured a likeness of Roberts holding a sword in his right hand. The figure stood over two skulls, with the initials "ABH" and "AMH." The initials stood for "A Barbadian's Head" and "A Martinician's Head." Roberts was partially successful: He captured Martinique's governor around 1720 and hanged him from his ship's mast.

liked the cheaper goods even though they didn't like Blackbeard's violent ways. Blackbeard bribed authorities who ruled over the colonies (which later became the United States) into allowing him to operate. But in 1718, the governor of the colony of Virginia grew tired of Blackbeard's lawlessness.

While Blackbeard partied with other pirates aboard his ship in his hideaway off the coast, the governor sent naval ships to capture him. Boxed in by naval ships on one side and shallow sandbars on the other, the pirates fought back. After intense fighting, Blackbeard and his crew rushed aboard a naval ship, thinking the crew members were dead. Instead, the crew was waiting below deck for a hand-to-hand fight. They captured Blackbeard and killed him, then cut off his head, his braided beard still intact, and hung it from the ship as a warning to other pirates.

Not all pirates met such a violent end, and some were even honored by kings and queens. One such pirate was Sir Henry Morgan (c. 1635–88), who was born in Wales. Morgan fought for the British navy against the Spanish in the Caribbean and eventually led a number of successful raids upon Spanish ports. By 1665, Morgan was extremely wealthy and had moved his entire family to Jamaica, which was controlled by the British. There, he was named colonel of the port militia, which was responsible for the defense of the island.

Morgan later became an admiral in the British navy and traveled throughout the Caribbean and the Gulf of Mexico, leading raids upon Spanish-held Cuba and Panama and coming back with thousands of pieces of gold and silver. In December 1670 and January 1671, Morgan led a raid on the Spanish in Panama, but because this attack came at a time when England and Spain had declared a truce, Morgan was arrested and sent back to England. Soon, though, King Charles II (1630–85) of England was seeking Morgan's advice on Jamaican affairs and eventually sent him back to the island to serve as deputy governor. There, Morgan lived off his riches and acted like many a pirate, drinking rum and gambling in Jamaica's ports. He died of disease in 1688.

Occasionally, wives and mistresses accompanied pirates on the seas. At times, women even served as pirates themselves. Anne Bonny (c. 1700s) and Mary Read (c. 1690–1720) were perhaps the most famous female pirates of all time. Although Bonny was born in Ireland and Read in nearby England, the two actually crossed paths for the first time in the Caribbean. Bonny, who had married a seaman while living in South Carolina, moved

Anne Bonny

to the Bahamas with her husband, but while there, she fell in love with the pirate "Calico Jack" Rackham (c. 1700s). She left her husband, took up with Calico Jack, and joined his pirate crew, often wearing a male disguise.

Meanwhile, Read joined the British military, married a soldier, and opened a hotel in England. When her husband died, she decided to make a new life for herself and boarded a Dutch ship bound for the West Indies, never revealing herself to be a woman. While traveling, her ship was attacked by Rackham and Bonny's pirate crew. Read was forced to become a pirate, and soon Rackham, Bonny, and Read were sailing together around the Caribbean, attacking ships and capturing treasures. In 1720, their ship was captured by the British navy, which was seeking to eliminate piracy on the sea. The pirates were sentenced to death. However, neither Bonny nor

Read, who were both pregnant at the time of their capture, was executed. Read died in prison of fever, while Bonny gave birth in prison, received a **pardon** from the British, and was released but never heard from again.

In the Asian seas, female pirate Cheng I Sao (1785–1844) of China became one of the most powerful pirate leaders of all time. After her husband died in 1807, Cheng took over an impressive pirate fleet operating along the coast of China and Malaysia, with upwards of 1,500 ships and 70,000 to 80,000 men and women under her command. These types of large pirate fleets were not uncommon in the Asian seas. Those who ruled over them did so with power similar to that of emperors and kings. Cheng was known for her cruelty. Often, her method of punishment involved beheading.

Meaty Meal

A favorite meal of pirates when food was available was called salmagundi. This dish featured a variety of meats—turtle, fish, chicken, pig, cow, duck, and pigeon. The pirates roasted all the meats and soaked them for several hours in a red wine sauce. They seasoned the meat with pepper, salt, garlic, oil, and vinegar. On the side, they might eat cabbage, anchovies, mangoes, onions, and grapes. Pirates rarely ate this well and spent long weeks aboard ships eating nothing but biscuits. These rare, delicious feasts were to be savored!

The Chinese government ruthlessly pursued Cheng, but its ships proved to be no match for her fleet, known as the Red Flag Fleet. The Chinese government suffered many casualties in fighting her crew. But by 1810, arguments within Cheng's pirate ranks led to a weakened organization. In addition, the Chinese government tried to eliminate piracy and regain control of coastal waters by offering pardons to pirates who surrendered. Cheng took the Chinese up on this offer, surrendering and obtaining pardons for herself and her crew members. Reportedly, she operated a gambling house and brothel in Canton, China, until her death in 1844.

End of an Era

By the 1800s, the golden age of piracy was on the decline. European nations that had wrestled for control of ocean waters and new lands were growing more stable, so they relied less on privateers to **ransack** enemy ships. In addition, the technology was in place to create stronger and faster ships, equipped with the most advanced weapons to defend themselves. Pirate crews who worked for themselves and wasted all of their money while at port could not afford such weapons and were unable to stand up against the new, well-organized navies.

OPPOSITE: As technology changed and governments evolved, the sun set on the golden age of piracy.

OPPOSITE Barbary pirates terrorized merchant ships for hundreds of years.

By the late 18th and early 19th centuries, governments also began to clamp down on illicit trade. Stories of Blackbeard's terrible demise made their rounds and encouraged Atlantic pirates to give up their lawless ways. In the Mediterranean Sea, the 1816 bombardment of the northern African port of Algiers put a stop to **Barbary** piracy. Barbary ship captains had a practice of enslaving Christians and forcing them into piracy, which the British and Dutch abhorred. When the government in Algiers, which had grown rich from the Barbary pirate trade, refused to put an end to piracy, the British and Dutch brought in a fleet of ships and attacked the Barbary pirates with gunfire. In the end, Algerian officials agreed to sign a treaty. They sensed that the British and Dutch would surely win if skirmishes continued. The treaty released Christian prisoners and effectively ended piracy in the Mediterranean.

The 1816 bombardment of Algiers

In the Asian seas, the British navy successfully defeated pirate fleets in the 19th and early 20th centuries. One of the last big battles between a government and pirates occurred in 1849 off the coast of Vietnam. The British and the Chinese sought to eliminate piracy, which was hurting the **opium** drug trade between Asia and Europe. The fleet of Shap'ng-Tsai (c. 1800s–1849), one of the last great Asian pirates, was no match for the firepower of the large British warships and was destroyed.

Despite the fact that the exploits of pirates had largely come to an end by the latter part of the 19th century, piracy still exists today. Modern-day pirates use technology, intercepting e-mails or telephone calls to locate ships and arming themselves with deadly assault rifles and rocket-propelled grenades. Today's pirates attack all types of ships—local merchant ships, international tankers, and even passenger cruise ships. Relying upon sneakiness, surprise, and disguise, pirates seek ways to board target ships. Gaining control of a ship today can be easy because large ships use few crew members.

Today's piracy most often occurs in the South China Sea, the Red Sea, and the waters off South America and

India. The most dangerous route is said to be the Strait of Malacca near Indonesia, Singapore, and Malaysia, where 97 pirate attacks were recorded in 2020. Modern pirates aren't interested in taking a ship's entire stock, as were their predecessors. Today's massive ships hold gigantic amounts of cargo, which would be impossible for pirates to grab. Instead, the sea robbers take the crew members' and passengers' personal possessions or money kept aboard the ship. Many of today's pirates engage in illegal trade, avoiding government taxes by smuggling goods—such as tobacco and alcohol—into countries, including the United States. They also sneak in drugs and people who are part of the modern-day slave trade.

Sometimes, though, modern pirates don't steal or smuggle anything. Instead, they may attack a ship for political reasons or hold people aboard the ship hostage

A small U.S. Navy crew delivers assistance to a fishing ship previously seized by Somali pirates.

Pirate Haven

By the 18th century, when governments actively sought to root out pirates, some pirates took refuge on the island of Madagascar off the eastern coast of Africa. Madagascar was not ruled by any country, so pirates took it over and made it their own. The island was situated in an important location in the vast trade network between Africa and Southeast Asia. It was full of riches, and its isolated nooks and inlets made it an inviting place for pirates. Madagascar soon became a popular haven for pirates who wanted to retire from the seas but did not want to return home.

and make demands for their release. Such was the case in the 2009 hijacking of the Danish/U.S. cargo ship *Maersk Alabama* off the coast of Somalia. Four teenage Somali pirates, armed with assault rifles, boarded the ship and tried to take control, eventually taking Captain Richard Phillips hostage and fleeing in a lifeboat. The ordeal and subsequent rescue of Phillips was featured in the 2013 film *Captain Phillips,* starring actor Tom Hanks.

Although today's pirates are often dismissed as dangerous renegades, the world takes a

romanticized view of the pirates of long ago. Fictional accounts of pirates tend to be lighthearted and over the top. The more serious crimes of pirates are often overlooked in fiction, as are the true, horrifying conditions in which pirates lived.

One of the most popular pirate tales of all time is Robert Louis Stevenson's book *Treasure Island*, published in 1883. Featuring the **buccaneer** pirate Long John Silver and his quest to bury his treasure of gold, the book gave rise to the idea that "X marks the spot" indicates buried treasure on a map. Another enduring pirate story is the beloved *Peter Pan* by J. M. Barrie, in which the notorious Captain Hook kidnaps Peter Pan's friend Wendy.

Pirates have also graced the silver screen since the invention of the moving picture. Stories of **swashbuckling** pirates became popular in the early days of film, as their daring adventures and swordfights made for dramatic scenes. In 1926, American actor Douglas Fairbanks starred in *The Black Pirate*, and in 1935's *Captain Blood*, Errol Flynn played a British surgeon-turned-pirate who robs from the rich and gives to the poor. Nearly a century later, pirate movies are still popular. The most famous pirate on screen of late is Johnny Depp's portrayal of Captain Jack Sparrow, a colorful Caribbean pirate, in the 2003 movie *Pirates of the Caribbean: The Curse of the Black Pearl* and its 2006, 2007, 2011, and 2017 sequels.

The pirates of the golden Age of Sail remind us of a time when vast riches crisscrossed the world aboard ships and formed the beginnings of a global trade network now firmly in place. Pirates of the past aimed to terrorize the innocent, and their hostile ways can still be found among pirates who operate on the seas today. Yet, while pirates leave little to emulate, their stories continue to fascinate.

Sunken Legend

Pirate Sam Bellamy (c. 1689–1717) was a notorious sea robber until his ship, the *Whydah*, sank in the Atlantic Ocean near Cape Cod in 1717 during a violent storm. As a result, 144 men died, and down went treasure from more than 50 captured ships. The sinking remained a legend and the ship's whereabouts unknown until underwater explorer Barry Clifford tracked it down in 1984. Since then, more than 200,000 artifacts, including cannons and silver, have been recovered from the ship. The *Whydah* is the world's only authenticated pirate shipwreck.

Selected Bibliography

Alberge, Dalya. "Explorers Unlock the Mystery of 'Pirate King' Henry Avery Who Vanished after Huge Heist at Sea." *The Guardian*. March 30, 2024. https://www.theguardian.com/world/2024/mar/30/explorers-unlock-the-mystery-of-pirate-king-henry-avery-who-vanished-after-huge-heist-at-sea.

Kingsley, Sean. "Who Were the Real Pirates of the Caribbean?" *Smithsonian* magazine. May 15, 2024. https://www.smithsonianmag.com/history/who-were-the-real-pirates-of-the-caribbean-180984343.

Kingsley, Sean, and Rex Cowan. *The Pirate King: The Strange Adventures of Henry Avery and the Birth of the Golden Age of Piracy.* New York: Pegasus Books, 2024.

Lehr, Peter. *Pirates: A New History, from Vikings to Somali Raiders*. New Haven, Conn.: Yale University Press, 2019.

Moore, Michael Scott. *The Desert and the Sea: 977 Days Captive on the Somali Pirate Coast*. New York: Harper Wave, 2018.

Phillips, Richard, with Stephan Talty. *A Captain's Duty: Somali Pirates, Navy Seals, and Dangerous Days at Sea.* New York: Hyperion, 2010.

Glossary

Age of Discovery	the period from the early 1400s to the 1600s when European explorers traveled worldwide by sea
Age of Sail	the period from the 1500s to the mid-1800s when sailing boats dominated international waters and played an important role in global trade
astrolabe	a navigational tool used to measure the height of heavenly bodies from the horizon; from this, sailors could deduce their location and speed
Barbary	a term used until the 19th century to describe the area of northern Africa along the coast of the Mediterranean Sea; it included the coastal areas of modern-day Morocco, Algeria, Tunisia, and Libya
booty	the treasure pirates took by force from a ship, usually in a daring manner
buccaneer	a name originally given to inhabitants of the West Indies and later applied to pirates who traveled in that area of the Caribbean
democratic	relating to a system of government that favors equality and decision-making by the people or by representatives elected by the people
fleet	a group of ships used by one crew; famous pirates often had dozens or even hundreds of ships in a fleet

hand grenade a small bomb thrown by hand; a grenade is designed to explode upon impact, injuring or killing its target

merchant ship a vessel that carries goods such as gems, silver, gold, and food across the ocean to be sold at different markets

opium a narcotic drug taken from the dried juice of poppy seeds

pardon a release granted by someone, usually a government official, that excuses a person from further punishment

plunder to take goods and treasures by force or violence, often causing damage to the surroundings in the process

ransack to search thoroughly and often steal some things, destroying almost everything in sight in the process

Roman Empire the vast territory, centered in Rome, that included most of modern-day Europe, Asia Minor, and northern Africa; the Roman Empire lasted from the eighth century BCE to 476 CE

romanticized thought of in a positive, romantic, or idealized way, often ignoring negative qualities

scurvy a disease common aboard ships, caused by a lack of fresh fruits and vegetables, which contain vitamin C; scurvy is marked by swollen, bleeding gums and easy bruising

spyglass a small, handheld telescope or set of lenses used to spot things that are far away

swashbuckling acting in a proud, carefree manner; having a swaggering walk, especially by an explorer or adventurer

Websites

The Golden Age of Piracy

https://www.rmg.co.uk/stories/topics/golden-age-piracy
Discover more about pirates of the 17th and early 18th centuries.

International Talk Like a Pirate Day

https://talklikeapirate.com
Have fun with a pirate name generator, pirate songs, and more.

Pirate Interdiction and the U.S. Navy

https://www.history.navy.mil/browse-by-topic/wars-conflicts-and-operations/pirate-interdiction.html
Explore the long history of conflicts between pirates and U.S. naval forces.

Index